Andrew McDeere

Raising Nigerian Dwarf Goats

A complete Guide to Raising, Breeding, Keeping and Take Care of your Nigerian Dwarf Goats

Copyright © 2020 publishing.

All rights reserved.

Author: Andrew McDeere

No part of this publication may be reproduced, distributed or transmitted in any form or by any means, including photocopying recording or other electronic or mechanical methods or by any information storage and retrieval system without the prior written permission of the publisher, except in the case of brief quotation embodies in critical reviews and certain other non-commercial uses permitted by copyright law.

Table of Contents

Nigerian Dwarf Goats 4

Nigerian Dwarf Goats Are Easy Keepers That Really Produce 19

Nigerian Dwarf goats on farm 28

How To Get Started With Nigerian Dwarf Goats 34

Care of Nigerian Dwarf goats 41

Feeding Your Nigerian Dwarf 44

Facts Of The Nigerian Dwarf Goat Breed 55

Small Goat Breeds 71

Nigerian Dwarf Goats: General Care 75

Getting Nigerian Dwarf Goats? Identify Your Purpose 89

Things to Know About Raising Goats as Pets 100

Frequently Asked Questions About Nigerian Goat 106

Nigerian Dwarf Goats

Have you ever thought about having a pet goat? Nigerian Dwarf goats are of West African origin. They are small dairy goats standing less than two feet high. They have calm, uniform, kind, gentle and playful personalities. They are easily recyclable and can also walk on a leash. These features make them adorable companions or pets, especially for young children, people with disabilities and the elderly.

They are available in several colors, including: white, black, red, cream and patterns. Some have a white "enamel" on their ears and blue eyes. The nose is straight and the ears are erect. Their peeling is gentle with short and medium hair.

Nigerian goats peacefully share pastures with other animals such as cattle, horses, llamas and donkeys. They help improve pastures by eating weeds, bushes and ivy that other animals will not eat.

These breeds of goats need ventilation for optimal health. They should not be kept in hermetic buildings. Their feathers should be clean, without drafts, without pests such as flies and rodents. Fresh straw or hay should be used for bedding. Some owners have used a large niche, or two, with toys included for their playful character. They can survive in many places, including places in cold climates, such as Ontario, Canada.

Your hooves should be cut, regularly, every four to eight weeks, and vaccinations should be examined with a veterinarian. Deworming is also important for maintaining good health. Good hay, pastures and fresh water, in clean containers, should always be available. With proper care of the goat, its average duration can vary from fifteen to twenty years!

They can produce a range of one to eight pounds of milk per day. Its milk has a sweet taste due to a high fat content and is excellent for soap making.

Nigerian Dwarf goats can breed all year round. They can have several children at once, with triplets and quads. They are excellent mothers and take care of their children, if allowed.

If you have never considered owning Nigerian Dwarf goats as pets, you can reconsider. These miniature goats are not only friendly and easy to train, but are ideal for participating in goat shows.

One of the twenty best events across the Southeastern United States is called the goat festival, music and more. This festival will take place in Lewisburg, Tennessee from October 8-10, 2010. This festival is not just a small event, it has grown from a few thousand in its first year to more than 17,000 participants from Fifty United States and four countries. At this festival you will find a variety of goat shows, arts and crafts, goat milk soap, activities for children and adults, goat milk soap producers, a variety of goat milk products, live music groups, Motown Music has gospel Songs, Artists and country music artists.

Most of us like to have cute animals. Pet goats like Nigeria's dwarf goats come from West Africa, just two feet long, a small version of a goat you can take care of at home. Animal lovers will definitely appreciate having them as cute pets because they have quiet, imperturbable, adorable, tender and playful characters. They can be easily trained and able to walk on a restriction. They are easy to be with especially with children and seniors because they are adorable.

There are assortment varieties in terms of colors such as white, black, cream, red and patterns. There are also special origins where they have white enamel on their ears and some have blue eyes that make them even more adorable.

Nigerian goats grazing with other animals such as cows and donkeys. They eat weeds that help maintain the crop because other animals refuse to eat them. For optimal health, your feathers should be well ventilated. Your pens should be kept clean and there should be good air circulation in the building. To survive in different places, they maintain a playful character. They used a large kennel with toys included. The hoof of the Nigerian dwarf goat must be well cut and the vaccines must be properly controlled. To maintain good health, goats need to be dewormed. If the goats are well groomed, they can live up to twenty years.

There are many positive results of proper care, can provide a range of one to eight pounds of milk per day, and the milk will produce a delicate flavor and excellent to make soap.

Cute animals like Nigerian Dwarf goats can reproduce all year round and can have several babies at any time. So consider owning Nigerian Dwarf goats, they are really cute pets because they are friendly, easy to train and can be a great artist.

Nigerian Dwarf Goat Coloring

Color is one of the factors that make Nigerian Dwarf breeding so popular. You can never be sure of the color of children until they are born; even then, you can not be sure because many times their color can change. The main color families are black, chocolate and gold with virtually every imaginable combination produced. You can dye Dalmatians, painted prints, tricolor or simply elegant shades of solid jet black, white, chocolate or gold. Crown skin patterns are also common, described by contrasting facial stripes, a "coat" around the shoulders with a coordinated dorsal band and marks on the legs. Brown eyes are the most common; however, Blue Eyes from China are also possible.

About Nigerian Dwarfs

The Nigerian Dwarf is a miniature goat of West African origin. Nigerian Dwarf goats enjoy greater popularity due to their small size and colorful markings. Their small size means that they do not require as much space as their large goat slags counterparts. Their friendly and friendly personalities make them good pets and easy to handle, even small children can feel comfortable with these little goats. Nigerian dwarfs are still considered "rare" for the conservation of American beef breeds. American. The Ministry of Agriculture also approved the Dwarves of Nigeria as dairy goats for livestock, making them suitable for 4-h and FFA youth projects.

Health

Dwarf goats, like all other breeds, need basic care for good health and a long life. Hooves should be cut regularly, approximately every 4-8 weeks or as needed. A properly cut helmet must have the same shape as a goat helmet. Vaccination against tetanus and enterotoxemia of Type C and D are the basic types. Consult your veterinarian for more information or other vaccines recommended for your area. The worm should be done twice or more a year. Your veterinarian can suggest a good program for the needs of your herd.

Nigerian Dwarf Shows

Dwarf shows are gaining popularity and becoming more and more available. The shows are fun, educational and a great way to meet other breeders. They are a place to sell their goats or get higher stocks for breeding. View information or event can be obtained by ndga, local goat clubs and organizations. Can't find a show in your area? Talk to your district representative about organizing a show yourself. It is not so difficult and very fun!

Temperament

Dwarf goats are sweet and adorable. Their calmness, uniform temperament and attractive personalities make them suitable companions for everyone, including small children in 4-h or FFA. Breeders of other types of goats find that their dwarfs mix with the rest of their herd and do not need special parts; only a suitable fence to contain them due to their small size. Many Nigerian Dwarf goats peacefully share pastures with other animals such as cattle, horses, llamas and donkeys. In fact, they will often improve a grass by removing shrubs, weeds, poisonous Ivy, yellow Thistle and wild berry bushes that other cattle will not eat.

Accomodation

Goats should be stored in clean enclosures, free of moisture, drafts and pests such as flies and rodents. They also require appropriate fences due to their small size. Nigerian Dwarf goats should not be housed in hermetic buildings; they need ventilation for optimal health. For one or a few goats, many owners of a large kennel or two does the job. Feathers or houses should be cleaned with fresh hay or straw for bed linen. Many owners find that providing "toys" for their goats provides them with hours of Goat Entertainment. Tree stumps, rocks or cable reels are ideal for "King of the mountain" games and jumps. Be sure to keep them away from the fence unless you want Renegade escapes your loose pack in your neighborhood!

Feeding of Dwarf Goats

Most breeders feed 12-18% goat protein or milk ration. It should not contain urea as it is toxic to goats. Many breeders give less grain if grass and candle are available. Hay or grass should always be available for free. Fresh water in clean containers should also be available at all times.

Parental

Dwarf goats breed all year round. Many breeders breed three times in two years, giving the deer at least a six-month break. Of course, this is a personal choice for all breeders. The gestation period for a deer is from 145 to 153 days. For the most part, Nigerian dwarfs are an abundant breed with little joke problems. New children on average about 2 pounds. at birth, but they grow rapidly. Watch out for those little

Bucks! It is known that these little men reproduce and are fertile from 7 weeks. Be sure to wean fa and dollars separately so that this does not happen.

It can be served at 7-8 months of age if they have reached a good size. Some breeders prefer to wait until they are at least 1 year or more. The dwarf can have several children at once, 3 and 4 are common, and sometimes even 6. Dwarfs are usually excellent mothers who can take care of their children if they are allowed to raise their children. They can also provide an incredible amount of milk for your height if you decide to want to be delicious goat milk.

The coins can be used to serve 3-month-olds and easily when they are 7 or 8 months old. Dwarf dollars are vigorous breeders, but they are soft enough to be used for manual breeding or pasture breeding. Both methods are used successfully.

"Miniature dairy goats"

A range of Nigerian health dwarfs can produce a surprising amount of sweet milk due to its small size: up to three-quarters of a gallon per day. In addition, Nigerian Dwarf milk is richer in butter fat (6-10%) and richer in protein than most other dairy goat breeds. Not all Nigerian dwarf owners raise their goats for milk. Some simply enjoy the fun and the company these little goats bring into our lives.

Nigerian Dwarf Goats Are Easy Keepers That Really Produce

Notwithstanding their little size, Nigerian Dwarf goats produce a noteworthy measure of milk (on normal 2-3 pounds every day, and some convey twice that sum once in a while). Goat's milk has numerous medical advantages (more edible calcium, less calories and a metabolic structure that is retained all the more viably in the human body, to give some examples).

Numerous individuals are shocked that the little Nigerian dwarf is really a practical dairy goat. These bountiful little animals offer a great deal of milk, but at the same time are shockingly low upkeep. Being little, they need less food than most enormous dairy breeds.

It is likewise a solid variety that flourishes in hot and dry atmospheres, for example, the African deserts, from which it slips, in cold and sticky territories, for example, the horticultural grounds of Northern Ontario. They arrive in a wonderful assortment of hues and some are additionally designed, with an exceptional shading and stamps like those of the American Paint Horse.

These goats are amazingly delicate, brilliant and are superb animals. Numerous families keep them to help with garden the board, as they will chomp grass, weeds, little undesirable tree shoots and irritating nursery plants without annihilating the scene or biting over the cladding or wall.

With a fragile development and non-forceful nature, they don't fit the generalization of goats that devour everything and anything in their path (although this is actually somewhat misrepresented with regards to most assortments).

The NDG can without much of a stretch figure out how to stroll on a rope and a few proprietors even prepared them to break (not that numerous goat proprietors need indoor cows)! Do and wethers (emasculated men) make the best augmentations to household family life; dollars, while still delicate and well disposed, discharge something of a horrible smell.

Milk and dairy items, including yogurt and cheddar, just as goat milk cleanser, can be handily arranged at home. You needn't bother with much space to joyfully keep a few these clever animals; a little yard.08 section of land or works fine. It is prescribed to keep at least two goats, except if you have them with different animals (ponies, cows or delicate canines) so they are not taken off alone.

You can undoubtedly bring in some cash by selling your goat items! For business people who need to invest more energy into selling their family unit things, you can likewise set up an online site. Having a dependable broadband Internet administration with adequate data transmission to take client orders is fundamental in the event that you need to draw in and keep steadfast clients. On the off chance that you live in a rustic domain, satellite INTERNET gives all of you the highlights of online clients in urban territories.

In the event that you start a business, it is prescribed to offer just cleansers to non-nearby shoppers who spot orders on the web. Indeed, even with protected bundling, it tends to be hard to dispatch palatable dairy items and keep them in their best condition on appearance.

On the off chance that you have done all your examination and conclude that a goat is in your future, the Nigerian Dwarf is unquestionably worth considering! On the off chance that you choose to make your ND goat a colleague or a family pet, you will appreciate the sweet character and fun loving character of this awesome minimal pet.

Contrast BETWEEN NIGERIAN DWARF GOATS AND NIGERIAN PYGMY GOATS

In the event that you think the historical backdrop of the variety and the American history of Nigerian Dwarf goats and dwarf goats appear to be amazingly comparable, you are correct. The two varieties are fundamentally the same as in starting point and employments.

The fundamental distinction between Nigerian Dwarf goats and Nigerian dwarf goats is their body extents. Nigerian dwarf goats will in general have full barrel-formed bodies, much like ordinary measured goats with short legs. Nigerian Dwarf goats, then again, are all the more carefully proportioned, as littler typical measured goats.

There are likewise two distinctive variety guidelines for Nigerian Dwarf goats, contingent upon whether they are utilized as dairy goats or not.

Nigerian Dwarf Goat United States

Like all goats, Nigerian Dwarf goats are valuable in fields to wipe out weeds and undergrowth and improve grass for eating animals. Their mild nature makes them simple to keep with brushing animals and doesn't require uncommon consideration or quarters

As a reproducing breed, they are qualified for 4H ventures. they are little and simple to oversee, making them an extraordinary contender for kids' activities, for example, 4H and FFA

They give a decent measure of value milk for their size. Their milk is incredibly high in fat, perfect for goat cheeses, creams and other goat items

They are polyestrous and can be developed throughout the entire year, so few females can keep up a normal flexibly of milk

They normally produce 3-5 kids for every joke, so a little group can develop rapidly, or youngsters and wethers can be sold as pets

They are exceptionally mainstream essentially as pets and pets, and their little size makes them simple to live in much more urban situations.

To put it plainly, for individuals who as of now have a little or enormous ranch, it is anything but difficult to just include Nigerian Dwarf goats and make the most of their various advantages. Also, for the individuals who are searching for a fun, amicable and beguiling pet, a Nigerian Dwarf Goat doesn't require substantially more space or care than most types of canines.

Nigerian Dwarf goats on farm

We pick Nigerian Dwarf goats for our ranch for a few reasons. As fledgling goats-and homesteading all in all - we figured the smaller size of the variety would be simpler for us learners to deal with.

We visited a Nigerian Dwarf ranch before passing by to give us a thought of the animal and were excited with its inviting character, sensible size and guarantee of high-fat whey to help us on our way to independence. Look at all the advantages of keeping Nigerian Dwarf goats and check whether you can oppose adding them to your own cultivating life!

Friendship

Nigerian Dwarf goats are charming and benevolent, and seeing them kicking on a rich grass is the matter of rustic dreams. They are so basic goalkeepers, at the degree of personality, that they are regularly suggested for 4-H activities or FFA clubs. One can hope for something else than a time of friendship in light of the fact that a solid and all around prepped goat should live from 12 to 14 years.

MILK

One of the incredible attractions of this variety is the high fat substance of margarine, its milk. Dissimilar to 2% to 6% of different varieties, these liberal minimal ones produce inconceivably smooth milk with 6% to 10% butterfat.

Despite the fact that they can be little in size, a deer can create in excess of a liter of milk for each day! Obviously, it relies upon the measure of milk your kids permit to drink and the recurrence of milk, however they can be a reasonable choice for a family who wouldn't like to feel overpowered by a few liters of milk a day.

At last, These little rich animals can be raised throughout the entire year, which implies you can scale your goat's flatulating projects to ensure you have a solid gracefully of sweet milk lasting through the year.

MEAT

Execute the goats for meat may not be your fundamental objective in possessing goats, dwarf Nigerian, yet on the off chance that you locate that after a couple of periods of tricks, you have such a large number of Males or females, useless, giving a solid wellspring of your meat is an another preferred position that these animals offer in the organization. You can't get neighborhood meat, grass-took care of or reasonably reproduced contrasted with your animals!

Compost

Goat compost is a remarkable expansion to cultivate beds. Investigate this connection for more data on the best way to utilize these effectively dispersible granules. Try not to release it to squander, and your future harvests will remunerate you luxuriously.

Extra INCOME

These benevolent animals likewise make fabulous animals. You might have the option to murmur an auxiliary pay for the homestead by selling a few wethers or a portion of your extra youngsters.

Since Nigerian Dwarf goats are productive, regularly with three or even five kids one after another, this can be a great alternative!

How to Get Started With Nigerian Dwarf Goats

Getting your first Nigerian Dwarf goats should never be a rash buy. Cautious arrangement and cautious framework will guarantee you and your new goats a smooth progress and a cheerful and sound coexistence.

As a matter of first importance, you ought to consistently plan to have more than one goat. Like livestock, goats mull alone ... shouldn't something be said about amazingly vocal about his desolate discontent. Two is a decent beginning.

Convenience GOAT

On the off chance that your goats get the four-legged adaptation of the Taj Mahal or more rural homes, there are things that ought to consistently be set up.

As a matter of first importance, all goats need a spot to escape time. In contrast to sheep, goats don't care to be in the breeze and rain and can endure whenever left in unprotected components. On the off chance that you have a horse shelter on your property, it bodes well to construct an enclosure directly nearby and let the goats in and out if fundamental.

In winter, grown-up Nigerian Dwarf goats grow a thick, fluffy layer that can adequately shield them from the cold until they are dampened by solid breezes. Fa and its infants, in any case, may require some additional consideration in the event that you choose to raise goats in the colder winter months.

Some wellbeing experts decide to warm their horse shelters with lights despite the fact that there might be a danger of fire, while others give high temp water for the duration of the day to warm their group. Another alternative may be the rearing atmosphere of them guarantees that the lone kid during the hotter months permits them to abstain from kidding through the most noticeably terrible of winter.

CHILDCARE

Each time you make the infant, coincidentally, it is essential to secure kids during their youth. For the initial not many days or weeks, keep mother and child on spotless and dry straw in the stable while tying, and the infant gets his overly essential colostrum.

At that point leave them outside, as long as the climate is dry and warm. Some goat guardians prescribe permitting kids to go out just if the time is over 50 degrees Fahrenheit, while others depend on the correct impulse existing apart from everything else to get their kids out.

FENCING

Another component to consider is fencing. Goats (even little goats like these) need a great deal of room to meander, touch and play, and all goats, paying little heed to their race, are slick people some degree. There are numerous choices for fencing, all with its focal points and inconveniences.

Welded wire is reasonable, yet goats rub winter hide for spring, or dollars with the smell of it in the nose, they can drive the wires and flee.

Woven yarn is costly, however it appears to work very well much of the time with goats. Regardless of whether it's much more costly, you get the littlest squares you can, possibly 4 by 4 inches or littler. Nigerian Dwarf goats are littler than typical goats and require a "denser" fence to guarantee their security. Also, kids can without much of a stretch sneak through bigger squares and it is realized that Horned goats assault their heads; a deadly slip-up on the off chance that you don't discover them in time.

The electric link is additionally an alternative, yet you may locate that a few goats adhere to the link... what's more, some don't.

The most ideal alternative may wind up being a mix, on the off chance that you place an invigorated wooden fence with plaited wire or increment a current fence with an electrical cable.

Security of the predator

A third thought ought to be the security of predators. We know the agony of losing a goat to a wild canine excessively well on our cultivate and have since gotten a watchman creature to ensure we never endure such misfortunes again. Goats may have horns, yet they don't conflict with a homeless canine, coyote or wild feline.

A devoted attendant, for example, an uncommonly reared monitor canine, a jackass or a leaf, gives your goats a specific organization, however can guarantee that any ravenous predator is sent to run on a vacant stomach.

Care of Nigerian Dwarf goats

Thinking about Nigerian Dwarf goats can without much of a stretch become a piece of your day by day schedule, however it will take some effort to become acclimated to it from the start. With respect to the support of any animals, they need time each day! No more opportunity for hasty travel or seven days in length get-away, except if you have somebody you believe who can deal with your pets while you're away.

On our homestead, we don't consider it to be harm. We needed this life. In the event that you are new to the development of animals, notwithstanding, you may need to give yourself an opportunity to adjust to what you should be a few times each day.

Obviously, goats should be taken care of and watered each day. Creature wellbeing starts with great food and clean water. I composed a top to bottom article about the numerous choices there are with regards to taking care of goats.

You ought to likewise give your goats a decent glance at any rate once per day: check their hide for bothers, take a gander at their hooves to check whether they should be cut and their pets (since they love them and are delightful). This cautions you about something strange, yet in addition about acclimating goats to control them, which is something that will be required with regards to draining the reality. Draining is additionally a day by day task frequently two times every day!

Try not to be threatened by the clear outstanding burden, however. In the event that you are resolved to keep the goats and are prepared to gain from the experience and experienced darlings around you, the awards of offering life to these animals far exceed the underlying torments of development. The delight of their belare welcoming when you open the outbuilding in the first part of the day, the silliness of seeing infants hop off stumps and shakes with no fix, and the fulfillment of getting their own milk from their animals, are probably the most extravagant compensations as a byproduct of their endeavors.

Feeding Your Nigerian Dwarf

Nigerian smaller people are not even pioneering feeders. They appreciate the new and all around oversaw pastures as much as they taste the thick grass. On the off chance that a delightful rose or on the other hand they are given a newly planted youthful tree, they will likewise eat up it with excitement. Despite the fact that goats like to gather their own food, they get along impeccably when they are brought all their roughage and grain. Feed, indeed, ought to consistently be accessible, in any event, for goats who go through a large portion of their days taking care of.

The blend of spices vegetables cut before long makes a fine and flexible roughage. Every grown-up creature will eat from 1 to 2 pounds per day except if enhanced with wheat. Wheat, then again, is viewed as a dietary enhancement to keep up milk creation at the most significant level that every goat is hereditarily equipped for giving. Wheat supplementation may likewise be fundamental for pregnant ladies with the highlight of their time, from nursing moms and when grass or feed is constrained. A goat feed with 12% - 18% protein is a decent decision for draining Nigerian Dwarf goats. Ensure that the food doesn't contain urea as it is harmful to goats.

Since food needs change starting with one goat then onto the next and from season to prepare, with regards to concentrates, the ace's eye swells the goat. A normal of one kilogram of concentrate is generally adequate by and large. For a scope of dairy items, the brilliant principle is to take care of one kilogram to think for two fourth of a liter of milk delivered. Extra enhancements incorporate minor element salt in free structure. Numerous raisers offer preparing soft drink which is the type of a goat Rolaids. Minor element salt fills in as a wellspring of sodium chloride and other significant minerals, in spite of the fact that focus on the measure of copper in the enhancement, since Pygmy and midget goats are delicate to it. It offers rewards and dishes without salt. They will know how much your body needs.

Clean water is the least expensive however most significant supplement in a goat's eating routine and ought to be accessible consistently. Goats, being ruminants, need a liberal measure of water to keep up the maturation of the stomach. Youngsters need a great deal of water to make milk. All goats drink more water in the warm season.

What researchers have figured out how to take care of dwarf goats and why

Studies have demonstrated that dietary admission in Nigerian Dwarfs and dwarf goats builds richness and strength of kids. On the off chance that they get plentiful excellent sustenance, they have more advantageous kids and produce more milk. In an ongoing report, researchers found that a half increment in body weight during childbirth could be accomplished by improving nourishment. Higher protein levels and higher vitality admission

Milk creation and the quantity of sound conceived youngsters have expanded.

Different investigations in dwarf goats demonstrated that undeveloped organism endurance and early fetal advancement were not influenced by low protein consumption in a range. In any case, youngsters were underweight during childbirth, here and there underneath the endurance limit, which was straightforwardly identified with helpless sustenance of kids. A more elevated level of sustenance prompted more advantageous youngsters, and the mother had the option to deliver more milk.

Poisonous Plants- Protect Your Nigerian Dwarf

Your Nigerian goats or dwarf pygmies will eat weeds or poisonous plants and it is important to check your lot or grass before grazing them. Look at the photos below for some of the most common poisonous plants.

Locoweed reduces reproductive performance in male goats and affects almost all aspects of reproduction in women. Locoweed summer is harmful for Nigerian male and female dwarfs and pygmy goats.

Prevention of poisoning

Prevention of development and congenital malformations carried out by lupins, developed hemlock and Nicotian species. it can be obtained using a combination of management techniques. Some of them coordinate grazing times, modify breeding times, reduce plant population through herbicide treatment, manage grazing to maximize herbaceous cover and intermittent grazing. Veratrum belongs to the family of Liliaceae and is a poisonous plant.

Ponderosa pine needles and Slytherin cause abortions, while lupines, Hemlock veil and nicotian glauca plants cause congenital malformations. See the photos below for identification.

Ponderosa Pine

Snake seeds

Poisonous hemlock

Nicotiana glauca

Lupin

Symptoms of intoxication

Clinical signs of development are probably caused by the neurotoxic alkaloids of Cevanine present in most species of Veratrum. Control of Veratrum is relatively easy with herbicides. Keeping sheep and other animals from grazing, containing during the first trimester of pregnancy is another method.

The recognition that plants can have a significant impact on reproductive performance is relatively new and is not fully realized.

Other plants, trees and weeds to be careful:

Bracken greenery lodge causes aplastic weakness.

Bermudagrass, Johnsongrass, sorghum grub, cherry shrub, cherry tidbit and dark cherry can cause cyanide advancement.

Adelpha, Gossypol, Mountain Laurel, azaleas, rhododendrons, Leucothea, Lyonia, Japanese Badger, Horned rye and a few fescues can cause cardiovascular brokenness.

Tung oil tree, Buttercup, strap, bladder, Rattlebox, Tallow, Nightshades, passing beds and Pokeweed can cause gastrointestinal indications.

Aflatoxins, Cocklebur, Crotalaria, Lantana, mildew covered feed and blue green growth can cause liver disappointment.

Espresso Senna and grass shearer are known to cause muscle brokenness.

Dallisgrass, Bermudagrass, Carolina Jessamine, Phalaris, Buckeye, Hemlock water, Hemlock poison, zigadenusspp and White Snakeroot are ceased to cause sensory system manifestations.

Corn stalks, sorghum, millet, straw, Bermuda grass, wheat and amaranth are known to cause improvement from nitrates or nitrites.

Peppermint, yam, grub grass and nut feed are known to cause lung brokenness.

Facts of the Nigerian Dwarf Goat Breed

The Nigerian Dwarf is a smaller than expected goat of West African starting point. Their compliance is like that of huge types of dairy goats. Body parts are in a fair extent. The nose is straight. The ears are erect. The coat is delicate with short and medium hair. Any shading or shading plan is satisfactory, albeit silver agouti (roan) is viewed as a moderate deformity.

High contrast Dalmatian deer design

Nigerian Dwarf goat tallness

Most extreme perfect

17 "to 19 " is 22.6""

Dollars 19 "to 20" 23.6"

It is recommended that the perfect weight is around 75 pounds as per hes assessment data.

The creatures are precluded from the ring show for being curiously large by breed principles, wavy head, Roman nose and hanging ears or myatonia test. (This is related with goat blacking out.)

What you can discover on overshadow goat

The most widely recognized inquiry concerning diminutive person goats is: what is the distinction among them and dwarf goats? In spite of the fact that they have comparable causes, they are particular and unmistakable varieties. Dwarfs are reproduced to be "stocky" and overwhelming skeleton. Diminutive people are reared to have the length of the body and structure, in extent, of a dairy goat.

Smaller person goats are accessible in a few hues. The principle shading families are dark, chocolate and gold. Arbitrary white imprints are normal, as are spots and other shading mixes, for example, red, white, gold and dark. Shading is one of the enormous elements that makes reproducing midgets so famous. You can never make certain of the shade of kids until they are conceived; and still, after all that, you can not be certain in light of the fact that their shading can change.

Gold and white deer

What you can discover in predominate goats are overall quite pleasant. Domesticated animals Dollars are additionally handily overseen. They make awesome creatures and extraordinary creature ventures for the little ones of every 4-H.

Raisers of different kinds of goats find that their smaller people wed well with the remainder of their crowd and needn't bother with uncommon lodging, just reasonable fence to contain them because of their little size.

Diminutive person goats breed lasting through the year. Numerous reproducers raise their FA multiple times in two years, giving the deer a 6 months additional rest. This is, obviously, an individual decision for each raiser.

Highly contrasting Dalmatian deer design

Babies on normal around 2 pounds. during childbirth, however they develop quickly. They arrive at sexual development at an early age, so make certain to isolate the dollars and do it. These little kids had the option to grow up and be as rich youthful as 7 weeks.

They can be reproduced at the age of 7-8 months on the off chance that they have arrived at a decent size. A few reproducers want to hold up until they fear 1 year or more.

FA diminutive person can have a few kids on the double, 3 and 4 be normal and at times even 5.

Smaller people are normally acceptable moms and ready to deal with their youngsters, you should let them do the childhood of kids. They can likewise give a mind boggling measure of milk for their size. They can give from three to four kilograms for each day from 6 to 10% Fat.

What you can discover on predominate goat can be utilized for administration as youthful as 3 months and effectively when they are 7 or 8 months. Diminutive person dollars are overwhelming reproducers, yet they are sensitive enough to be utilized for hands animals or touching. The two techniques are utilized effectively.

Smaller person goats are movable in 3 registers. The American Goat Society (AGS), the International Dairy Goat Registry (IDGR) and the Canadian Goat Society (CGS). Smaller person shows are picking up notoriety and are progressively accessible. Most are authorized by AGS.

Highly contrasting Range

While the quantity of Nigerian diminutive people is still little (just 3500 creatures are enrolled in the US with AGS), they have an exceptionally splendid and productive future.

Convenience

Goats ought to in a perfect world have a dry, concealed safe house in moist, cold or hot atmospheres. It doesn't need to be exquisite, an enormous doghouse works incredible for men and doesn't dry out. A more extensive secured region ought to be given during joke time, giving the reality enough space to move around kidding without misleading newborn children. They ought not be housed in airtight structures; ventilation is fundamental for ideal wellbeing.

Numerous proprietors offer " toys " for goats to play. Tree stumps, rocks or enormous link reels are ideal for " King of the mountain " games and hops. Make certain to keep all the "toys" away from the fence to keep the entertainers from getting away. A decent fence is essential as a result of its little size.

Nigerian dairy goat rearing

Nigerian goats can raise lasting through the year. The growth time frame for a deer is from 145 to 153 days. Generally, Nigerian goats are a bountiful variety with little joke issues. New kids on normal around 2 pounds during childbirth, however they grow quickly. Watch out for those little Bucks! Twists can be fruitful at 7 years old weeks. Make certain to wean and isolate the dollars to assist you with maintaining a strategic distance from those from automatic rearing.

The coins can be utilized for upkeep made as youthful as 2 months and surely for the second are 4 months. Nigerian goat guys are vivacious reproducers, yet they are sensitive enough to be utilized for hand rearing (contained) or field reproducing where one male is accessible for a few entering the estrus. The two strategies are effectively utilized.

You can get up just 4 to 5 months or once you begin accelerating.

Most reproducers want to hold up until they are at any rate 1 year or more before rearing. The greatest misstep that any raiser can make is to expect that the olive trees can not effectively breed a deer at about two months or that the deer can not be reproduced at an early age

Nigerian can have a few kids on the double, 3 and 4 are more normal than singles or twins, with births of multiple times. Nigerian goats are frequently acceptable moms and can deal with their youngsters easily.

They can likewise give a mind blowing measure of milk for their size.

See the milk program page for more data on draining..

Feeding of Nigerian goats

The most significant realities to recollect when taking care of goats is recognizing what mineral inadequacies are in your general vicinity and the kind of food that is promptly accessible, either from the field, a supermarket, or legitimately from the maker. Get however much data as could be expected from the reproducer about the enhancements they give your goats on the off chance that they purchase from somebody neighborhood. We regularly purchase goats from reproducers in different states. Note that every area of the nation presents various difficulties in the administration of the group. What works for somebody in the wettest areas of the nation, won't work for desert atmospheres. In the event that there are no different reproducers in your general vicinity, converse with your augmentation office to discover what to finish.

Goats need grub to keep their rumen cheerful and sound. Numerous goat raisers utilize 12% - 18% goat protein or a part of milk. It ought not contain urea as it is poisonous to goats. Goats will eat toxic Ivy with no destructive impacts and evacuate any brush or blossom in the event that it gets an opportunity. There are plants that are lethal on the off chance that you eat even a leaf. Roughage or grass ought to consistently be given in plenitude. In a perfect world, great field and cruising are the best.

Freshwater ought to be accessible consistently.

Wheat, squashed beet mash, sunflower seed dark oil(head), molasses, a business grain/wheat, or a blend of one of them is regularly given to breastfeeding done to keep up the condition while bringing up their kids or giving milk to their table.

Health

Nigerian goats, similar to every single other variety, need essential consideration for good wellbeing and a long life. Stops up ought to be cut normally, around each 4 two months or all the more regularly if vital. An appropriately cut and shaped head protector ought to look like those of a goat cap. There are fundamental kinds of antibodies to control every year, sorts of lockjaw and R and D. It is ideal to counsel your nearby veterinarian for every other antibody suggested for your region. Some accomplished raisers can inoculate their own goats; New proprietors and reproducers ought to carry their goats to the nearby veterinarian for immunizations.

Drying out ought to be done a few times each year. Your veterinarian may propose uncommon enhancements, (for example, selenium), extra antibodies, and a Recommended program of dewormers and dewormers for your specific run, contingent upon the area and realized preventive wellbeing measures.

Detection of specific diseases

These diseases can wipe out your flock. It is better to buy goats from proven flocks. That said, there are healthy flocks that have no flavor. More information is available online. Washington Animal Disease and Diagnostic Lab (Waddl) is one of the laboratories used for testing. Find out that this laboratory is the closest to sending samples for testing.

How Much Do They Cost?

The normal expense for the enrolled reproducing stock is between $ 200 and $ 500 for every individual, contingent upon the area and accessibility. Some of the time a reproducer will sell a creature for less on the off chance that he purchases a bundle.

Goats with test family or milk creation records can bring more significant expenses.

Quality pet stocks frequently cost considerably less with wethers, emasculated guys, for the most part accessible for $ 50 to $ 100.

It is smarter to contact a few reproducers in your general vicinity and see their groups before making the primary buy.

All things considered, the main goats bought are generally founded on feeling, not rationale. Where all the recently settled measures for compliance, milk creation, and structure were tossed out of the window. This is extremely satisfactory and experienced raisers are regularly prey to these charming little animals that we as a whole love to such an extent.

Please, when you're dependent, there will be others like " you can't have one."

Small Goat Breeds

There are two kinds of scaled down goats in the United States, the dwarf goat and the Nigerian smaller person goat. Albeit Nigerian dwarf goats and diminutive person goats share normal hereditary qualities, ranchers have started to raise them specifically to improve attractive attributes.

Dwarf goats were reproduced for little strong bodies since they were viewed as meat creatures and the Nigerian Dwarf was chosen for its dairy qualities and milk creation. Since the two varieties were so resigned, they became well known increments to pet zoos, and this helped fuel their fascination as pets.

Contrasts and similitudes between Nigerian Dwarf goats and Pygmies

Dwarf and Nigerian Dwarf goats share the equivalent gregarious and amusing character that pulls in the Little Farmer and the beginner. Indeed, even the most aloof and genuine can't manage without "helping" watching goats play.

The Nigerian Dwarf has a dairy character, and the dwarf goat is powerful and smaller

Dwarf is actually a diminutive person meat goat from Africa. Be that as it may, most are reared and acknowledged as creature interests, pets and pets. The Nigerian midget additionally originated from Africa. It is a small dairy goat, with the run of the mill qualities of the milk of the standard dairy breeds. Both arrive at a similar normal tallness, generally 18 to 20 inches high with the limit of 22 inches and somewhat more for men. In any case, here the similitudes among Nigerian and dwarf predominate goats end.

Dwarf goats are substantial, wide and Square in appearance. They have a wide temple, a thick and short neck and a shorter body than the Nigerian smaller person. They are called cobby and smaller with a body boundary that is bodied, very much ripped and relatively more extensive than that of different varieties. Dwarf is a well disposed, cautious and Hardy goat who is exceptionally social with different creatures and individuals.

While Nigerian Dwarf goats share a similar tallness midpoints and social and adoring characters like those of the dwarf goat, their appearance radiates dairy character just as their bigger dairy partners.

Regardless of having a similar tallness as the dwarf goat, the body boundary, length and bone are altogether different. Their bones are a lot more slender, more articulated with slight skin, level ribs and long, rich necks that are completely joined with firmly tied shoulders. Being dairy creatures, their bosom framework and draining limit assume a significant job in their family and in the assessment of their quality.

Nigerian Dwarf Goats: General Care

Similarly, as with all goats, Nigerian Dwarf goats don't require a lot of care. They can blossom with straightforward lodging, new water and a decent eating routine for goats. Giving them a lot of room to nibble on the correct grass and the brush likewise makes it a superior crowd. We keep a little group of enlisted Nigerian Dwarf goats on our ranch here in western Georgia. We deal with our little property by intersection and turning the crowd. Transitory feeders worked with wire boards likewise help feed in winter.

These goats are very little when they develop, around 18-21 crawls on the shoulder. They're anything but difficult to tame and extraordinary around kids. We can walk and touch every one of our goats with no extraordinary consideration, aside from routine taking care of.

They are wonderful for a little ranch while staying little and submissive.

Nigerians are a dairy raise and can deliver a lot of significant milk. This current goat's milk can be utilized to make an assortment of purchaser things, for example, cheddar, cleanser and moisturizer. We prescribe this variety to all who are keen on getting a goat. We suggest that you visit the NDGA site for suggested raisers in your general vicinity.

Fundamental goat care

As you can discover with most goats, this variety doesn't require a lot of care.

Fabricate an appropriate Goat House

When arranging convenience for goats, neighborhood climate ought to be considered. Here in the south, a three-sided shed with a decent rooftop approves of our mellow winters, however its zone may require a more confined structure. Everything from a huge pet hotel to a little shed ought to do the stunt as indicated by your necessities.

The lodging ought to give enough space to your goats to move serenely, and they appear to like being off the ground. One significant thing to recall with regards to goats is that they are interested and snack everything. At the point when I state everything, I mean the world! Ensure you don't leave anything uncovered or free that wouldn't like to be eaten, or that could hurt your goats whenever eaten.

I think his chewing propensity is the explanation a few people think they'll eat anything, similar to jars! This isn't accurate, coincidentally, they appear to honestly eat everything while at the same time eating just food and plants.

Guard your goats on the grass

Guarantee the wellbeing of your goats by controlling their touching for harmful plants.

Fencing for your goats

Fencing can be a touch of baffling for the beginner rancher, yet with a little examination and practice, you'll run miles of fencing in a matter of moments. There are excesses of fencing choices and techniques to cover here, yet I can give some goat explicit tips:

Make your fence additional durable. Goats simply like Aries bounce and rub their hips against the fence, and if it's not sufficient, they won't stop until they wreck it.

Use locking entryways with extra security cuts. Goats are known slick people. By and by, he had endless entryways connected to be opened by a keen goat. Open entryways will in general get the entire group out!

The goats will eat or decimate anything inside the fence. They will likewise eat something outside the fence that they can reach. Never fence in plants or trees you need to keep, the goats will eat any plant they can reach. Many scene plants can be harmful. Likewise recollect that goats like to climb, rub and harm everything in their way.

Check your fields for toxic plants. When fabricating your new fence, make certain to check and expel any plants that might be harmful to your group, a rundown of poisonous plants can be found at the connection beneath.

Goats like to ascend and play the ruler of the mountain.

Goats like to ascend and play the ruler of the mountain.

Feed the goat simple

Making care of goats appropriately is as simple as giving a reasonable goat diet each day. Goats will improve when they can effectively benefit from a sheltered field.

Feed your goats appropriately

Taking care of goats appropriately is as simple as giving a decent goat diet each day. A healthfully reasonable goat food can be found at your nearby food and grain store. Utilize a charger to expand the force flexibly for legitimate conveyance. Your channel must have sufficient space for the whole crowd; in any case, just the most commanded goats will have the option to eat. While including feed can absolutely improve a goat's eating routine, appropriate goat sustenance is actually everything necessary.

With regards to a goat diet, feed is actually your principle intrigue. They like a delightful grass, where they can snack grass and weeds. Goats are commonly not extremely requesting with regards to scavenge. At the point when a crowd has enough grain, it is sufficient to take care of a limited quantity of goat consistently, with a nutrient enhancement.

Transportation of their goats

We weren't prepared when we purchased our initial two Nigerian Dwarf goats. We wound up holding the two goat infants on our lap for the short excursion back. From this first assortment of goats, we have improved our transportation gear.

Subsequent to attempting a couple of various arrangements, we settled on a huge confine. A pet hotel works incredible for moving littler goats, and the greater part of them additionally overlap for capacity. The main clear downside is that a pet hotel frequently have a dangerous floor. Goats have hooves, and elusive hooves and floors don't blend well. A basic arrangement is to put a thick layer of feed or straw on the floor of the pet hotel. Indeed, even a little floor covering functions admirably, as long as there are no removable parts.

Veterinary Assistance

While goats are typically very solid creatures, they require an infrequent visit to the veterinarian. We live in a country territory, with a prospering horticultural industry. In our area, finding a veterinarian to think about goats is very simple. Be that as it may, this may not be the situation in your general vicinity. Make certain to discover a veterinarian who will take an interest in your group before purchasing your goats. This will take out the pressure of attempting to locate an all around qualified goat specialist during a crisis circumstance.

All goats ought to get a progression of infusions CDT, which is an inoculation routine that shields them from normal illnesses. Counsel your veterinarian to decide the ideal time for these immunizations.

Locate a certified veterinarian

While goats are generally very strong creatures, they will here and there need a visit to the neighborhood veterinarian.

Goats are sufficiently simple

Similarly, as with most cows, Nigerian Dwarf goats are very simple to raise and keep up. They have just a couple of essential prerequisites and most goats are very troublesome. All things considered, a decent book about dashing is consistently ideal to have close by. I have a ton of books regarding the matter and, as I would see it, this book offers a greater amount of your cash. It covers numerous helpful themes in a simple to-follow position. I can't help thinking that I generally allude to the material he offers a few times each month.

Ensure and plan ahead by building a Secure Fence and satisfactory lodging before purchasing your first goats. Likewise, ensure you have a reasonable goat diet nearby.

Having a decent method to ship goats and a most loved vet on the speed dial will likewise make your life somewhat simpler. Following the above tips, you ought to have a slow change to the life of a goat shepherd.

Protecting your Nigerian Dwarf with CDT Vaccine

There are just a couple of sicknesses to fear, and most have immunizations to forestall them.

Dwarf goats are a solid gathering, and on the off chance that they are appropriately housed and taken care of, it is impossible that go down with a feared infection. On the off chance that you start with solid goats, odds are adequate for them to remain as such.

Goats are not prone to capitulate to pneumonia (a malady to which they are unequivocally delicate) if your house is very much ventilated and liberated from drafts. The sound box is doubly significant for infant kids.

Indulging the sickness, or enterotoxaemia, can mean the passing of a goat that depends on focus or chows down on spring grass without roughage breakfast. Immunization you can keep this from occurring.

Nonetheless, your goats ought not be influenced in the event that you limit access to fixation, feed the day by day apportion in two servings and serve your preferred feed. On the off chance that, then again, you utilize overabundance concentrate to invigorate milk creation, shield your goats from enterotoxaemia with a yearly inoculation and end of Clostridium porphyrinogens CD-pathogen, and most antibodies likewise incorporate lockjaw pathogen.

Worms can be an issue wherein goats store excrement in their lodging or container of water, or touch a similar field a seemingly endless amount of time after year. The arrangement here is to rehearse controlled brushing and compose the taking care of gear so as to debilitate fecal sullying. An intermittent fecal test (take at any rate twelve granules of new compost from your veterinarian) will reveal to you whether your goats gather worms, just as the sort of worm, so you can build up an avoidance program.

Long-haired goats can be prey to lice and other outside bugs. An assortment of bug spray splashes, plunges, powders, injectables and pour-one are accessible in rural stores and domesticated animals gracefully indexes (for dairy goats, get a dairy-endorsed bug spray).

Bosom contamination, or mastitis, is a danger to any nursing deer. The areola openings of a support don't close following draining the Cradle (both by the human hand and by a nursing child). In the event that the deer is in a wet and messy litter, microscopic organisms can move there through the openings. Disease can likewise follow a bosom injury. Keep the case clean and evacuate objects that can harm a bosom brimming with milk.

Hooves of a goat continually develop and need intermittent slices to keep the creature from getting faltering. Since the development rate differs with the individual and season, check the head protectors consistently. Utilizing an attachment cutting instrument or some sharp scissors, make a little cut at once until the base of every attachment is corresponding to the noticeable development rings.

Getting Nigerian Dwarf Goats? Identify Your Purpose

On the off chance that you are thinking about getting Nigerian Dwarf goats, at that point it is a smart thought to recognize your Nigerian Dwarf goats: distinguish your objective to get the goats first.

New proprietors of goats once in a while bounce and purchase such a large number of goats of an inappropriate sort or breed from the start. At that point they feel overpowered or baffled and wind up selling their goats as opposed to getting a charge out of them.

A superior methodology is to deliberately distinguish your principle objective of possessing goats. In this way, purchase a couple of great goats that satisfy this objective. This makes it conceivable to encourage the property of goats with less pressure. On the off chance that you have a fabulous time and make the most of your first goats, you can even now develop the crowd later.

While picking your first goats, there are various choices to make before getting them. To begin with, you have to choose which breed and sex of goat will meet your objectives (read on for help on this), and afterward whether they will be enlisted or unregistered (see enrolled or unregistered). On the off chance that you pick enlisted goats, it is essential to figure out how to peruse a dairy goat family (perceive how to peruse a dairy goat family). Choices should in this manner be made with respect to horns (see getting dairy goats-free, horned or chickened), taking care of (see acquiring dairy goats-prey or jug), draining (see manual or machine draining) and that's only the tip of the iceberg.

Mediavin

As a variety, Nigerian Dwarf goats (see here for a short story) become incredibly well known because of their little size, brilliant milk to taste, simplicity of care and agreeable character. The accompanying table, accumulated by the enlistment quantities of the American Dairy Goat Association (Adga), unmistakably represents how rapidly its ubiquity is developing. In only ten years they have gone from being generally obscure to the most well known dairy goat breed!

Nigerian diminutive person goat has become the most well known dairy goat breed as of late because of its numerous constructive properties. Nonetheless, it is essential to choose the correct sort of goat during the main determinations to live up to your desires. This generally implies beginning with does (female goats) or wethers (emasculated male goats) rather than bucks (unblemished male goats).

The principal thing to recollect is to begin little with great goats and keep it reasonable. At that point select the kind of goats that will satisfy your objectives. Regardless of whether you will likely have a pet, a reproducing venture, milk or conceivably raising goats, the accompanying data can assist you with concluding whether to pick Fa, Fa or perhaps dollars.

What amount?

Goats are animals of the group, so they need a buddy or become despondent. Along these lines, you will require a blend of in any event two except if the GOAT is planned as a friend of another animal varieties, similar to a pony or a fire.

Youthful clasping and wethers

PET

In the event that you need Nigerian Dwarf goats for pets, at that point purchasing a few youthful wethers bodes well. We don't think about the hormonal changes of brands or dollars. On the off chance that they are all around mingled, they make the goats gentler and cordial.

Indeed, even the best raisers have surplus wethers accessible every year, so they are truly moderate at about $ 100 each. Picking youngsters subsequent to weaning or taking care of a container, you can ensure they are associated as they grow up.

Youth training venture

On the off chance that goats are for a kid's rearing venture (4H or Future Farmer of America [FFA]), at that point purchase a range or doeling (youthful female goat) and on the off chance that it may be a decent method to begin. It relies upon the territory of the nation wherein you live and this record contains shows in your general vicinity. Numerous records do exclude shows for wethers (and there are not many shows for dollars), which makes it important to buy a range or doeling for a kid's raising undertaking. It is essential to check your region's records and rules before choosing what to purchase.

That you are considering purchasing a doeling and indeed, it is a smart thought to book them ahead of schedule as opposed to holding up until spring after the goat children are conceived in light of the fact that, around then, they are typically guaranteed to another purchaser. Each spring we get calls and messages from guardians searching for good quality kid raising undertakings, however we can't help them since the entirety of our goat kids are reserved. I recommend beginning the exploration in the fall, as this is when raisers regularly design and distribute their rearing projects.

MILK PRODUCTION

For the individuals who need the stunning milk delivered by Nigerian Dwarf goats, at that point purchasing two great quality doelings may bode well. By taking them home following weaning or taking care of the container, you can ensure they are very much associated as they develop. Also, you will have the opportunity to figure out how to deal with them before you have to get familiar with reproducing, cooling and draining.

Numerous quality doelings of major draining lines are reserved a long time before birth, so it is imperative to book ahead of time. The impediment of this methodology is that you should hold up in any event a year and most likely two as the doelings develop before you can get this imposing milk.

For the individuals who promptly need milk, one methodology is to purchase a deer in milk and its doeling. This gives you the advantage of quick milk without managing rearing and reviving right away. You will in any case need to figure out how to milk and care for the develop deer, yet you will value the advantages that go connected at the hip with raising your doeling. Finding a deer and doeling for selling wide draining lines can be troublesome, making it a procedure that can set aside effort to execute. Another methodology for the individuals who need milk is to purchase two realities in milk. Once more, you get prompt access to milk and finding an incredible more established quality is once in a while simpler than finding a more youthful stock. Numerous reproducers attempt to prepare for their most encouraging works and settle on the choice to isolate from certain people of yore. Regularly, these old ones are of incredible quality, yet since there is just a great deal of room on each ranch, some of them are sold.

BUCKS AFTER

I don't ordinarily suggest beginning with a dollar. Guys don't create milk, are bad pets, have an awful stench during the mating prepare and have appalling propensities (as indicated by people). Except if you're certain beyond a shadow of a doubt you need to raise Nigerian Dwarf goats, I wouldn't begin with a dollar. Regardless of whether you need to get into rearing, I encourage you to get an insecurity (flawless youthful male goat); so you can figure out how to cherish yourself as it develops and before it becomes "fragrant"."

There are numerous motivations to get goats other than those referenced above and without a doubt appropriate blends, wethers and even dollars. Whatever the reasons, it's typically best to begin with some great goats to keep it charming while you learn.

Things to Know About Raising Goats as Pets

You may have heard the phrase: "he really has my goat!"But what do you get when you get a goat? People think about getting the goat yard for many reasons: for milk, entertainment and companionship, or even to help keep some of the weeds cut. But before considering buying a goat, it is important to know how to keep them healthy and happy. Here are 10 things you need to know:

1. First, check the local laws.

Before you get a goat, make sure they are allowed where it lives. Check your city's regulations to make sure goats can be kept within your city limits and if there are restrictions regarding the size or weight of goats. Also, keep in mind that many cities regulate the proximity of animals to nearby homes or properties. Goats can also be very noisy, so before you adopt or buy one, make sure your neighbors are tolerant.

2. Mini vs Standard.

There are two sizes of goats, miniature breeds and standard sizes. Standard size breeds, such as Nubia or Alpine, weigh between 100 and 200 pounds. or more. Mini goats, such as pygmy and Nigerian dwarfs, tend to be more popular in urban areas due to the many local restrictions on the size and weight of goats (these smaller breeds tend to weigh 100 pounds. or not). If your garden is also miniature, make sure that the little goat or "child" goat you bring home does not grow as expected (goats are called Children, female goats are called facts, uncastrated males are called males and castrated males are called wethers).

3. Space to circulate.

Goats are active and playful. A miniature goat requires a minimum of approximately 135 square feet of Break Room space; a standard goat needs twice, with the area multiplied by the number of goats it has. Goats, large or small, need a yard that offers part of the sun and part of the shade and is protected from strong winds. Goats also need an attached, draughty shed or barn to cover, sleep, and protect themselves from predators and extreme temperatures. All windows in your enclosure should be higher than the head of the highest goat when standing on the hind legs. Otherwise, the windows should be covered with bars or screens so that a goat cannot pierce its head. Inside, a deer and its children need a 4 'x 9' stand for comfort. You will also need a place to safely store your food and a place to get rid of your dirty clothes.

4. A good fence is the key.

Goats like to rub fences, especially when evicted, or try to stick their heads through openings to eat a delicious item out of reach. Goats can also leave their pens. All this means that goats are difficult to resell. It should be reinforced with solid wood poles pushed deep into the ground and sealed with slats close enough to prevent their heads from slipping. It should also be large enough to hold them (4 'tall for mini goats; 5' tall for standard). If you use the thread, make sure it is strong enough that a goat cannot bend it or push it down to escape. Goats also like to chew wood, so if you have wooden fences, be prepared to replace them every few years.

Frequently Asked Questions About Nigerian Goat

WHAT IS THE SIZE OF NIGERIAN DWARF GOATS?

These little goats average only 17 inches to 19 inches and 19 inches to 21 inches per dollar and should average 75 pounds

HOW LONG DO NIGERIAN DWARF GOATS LIVE?

With proper care, Nigerian Dwarf goats can live between 12 and 14 years, except for any unforeseen illness.

HOW MUCH SPACE DO NIGERIAN DWARF GOATS NEED?

Nigerian dwarf goat shelters can be as basic as a large dog house, but to keep the goats happy, they need space to play. A fenced area of about 200 square feet per goat with a fence of at least four feet high should keep your Nigerian Dwarf goats happy.

WHAT DO NIGERIAN DWARF GOATS EAT?

Like most farm animals, their Nigerian Dwarf goats will eat hay, as long as it is in some kind of crib or canal so that it does not touch the ground. Goats will also feed and eat almost any growing plants they can reach. Make sure that the plants you eat are not dangerous and are not the ones you do not want to eat them.

Cereals are also an important supplement, especially for mothers and children.

HOW MUCH DO NIGERIAN DWARF GOATS COST?

The cost of Nigerian Dwarf goats depends on age, sex, pedigree and purpose. On average, you can expect to pay between $ 200 and $ 500 per goat. If you keep them as pets, there are good pets available for $ 50 to $100.

HOW LONG ARE NIGERIAN DWARF GOATS PREGNANT?

The gestation period of a Nigerian dwarf deer is from 145 to 150 days.

WHERE TO BUY NIGERIAN DWARF GOATS?

Registered breeders will make sure to get the healthiest goats, but local farms may also have good options.

HOW MANY NIGERIAN DWARF GOATS PER ACRE?

Per acre of grass for grazing, most sources recommend 1/10 of an acre per goat, making 10 goats per acre a reasonable number for grazing.

WHEN DO NIGERIAN DWARF GOATS STOP GROWING?

Nigerian Dwarf goats reach maturity in one year, unlike most standard-sized breeds that are not fully grown for about two years.

DO YOU HAVE TO MILK NIGERIAN DWARF GOATS?

The short answer is yes. The Nigerian Dwarf will continue to produce milk, whether he treats it or not, which can cause pain in the head and can cause mastitis, which is inflammation of the breast due to infection.

HOW MUCH MILK DOES A NIGERIAN DWARF GOAT PRODUCE?

A range of Nigerian dwarfs can produce two quarters of a gallon of milk per day, high-fat butter content of 6% to 10%.

WHEN TO RAISE NIGERIAN DWARF GOATS?

Although the hace can begin reproduction at 7 months, most sources discourage reproduction until the goats are fully grown within the year. There is no season for breeding Nigerian Dwarf goats, they breed all year round.

112

Kind reader,

Thank you very much. I hope you enjoyed the book.

Can I ask you a big favor?

I would be grateful if you would please take a few minutes to leave me a gold star on Amazon.

Thank you again for your support.

Andrew McDeere

www.ingramcontent.com/pod-product-compliance
Lightning Source LLC
Chambersburg PA
CBHW080502220526
45465CB00006B/2351